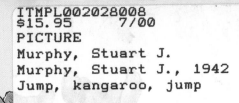

MathStart® **FRACTIONS**

JUMP, KANGAROO, JUMP!

by STUART J. MURPHY • illustrated by KEVIN O'MALLEY

HarperCollinsPublishers

LEVEL 3

To Phyllis Goldman—who is always ready to jump in with *winning* ideas

—S.J.M.

For Kate and Eliza

—K.O.

HarperCollins®, ☫®, and MathStart® are trademarks of HarperCollins Publishers Inc.
For more information about the MathStart series, please write to
HarperCollins Children's Books, 10 East 53rd Street, New York, NY 10022,
or visit our web site at http://www.harperchildrens.com.
Bugs incorporated in the MathStart series design were painted by Jon Buller.

JUMP, KANGAROO, JUMP!
Text copyright © 1999 by Stuart J. Murphy
Illustrations copyright © 1999 by Kevin O'Malley
Printed in the U.S.A. All rights reserved.

Library of Congress Cataloging-in-Publication Data
Murphy, Stuart J., date
    Jump, kangaroo, jump / by Stuart J. Murphy ; illustrated by Kevin O'Malley.
        p.   cm. — (MathStart)
    "Level 3."
    Summary: Kangaroo and his Australian animal friends divide themselves up into different groups for the various field
day events at camp.
    ISBN 0-06-027614-2. — ISBN 0-06-027615-0 (lib. bdg.) — ISBN 0-06-446721-X (pbk.)
    1. Division—Juvenile literature.   2. Fractions—Juvenile literature.   [1. Division.   2. Fractions.]
I. O'Malley, Kevin, 1961– ill.   II. Title.   III. Series.
QA115.M8714   1999   97-45814
513.2'14—dc21   CIP
                                                                                                       AC

1   2   3   4   5   6   7   8   9   10
❖
First Edition

"It's Field Day!" yelled Kangaroo.

There were twelve campers in all: a kookaburra, an emu, two platypuses, three koalas, four dingoes, and Kangaroo himself.

They warmed up as they waited for Ruby,
the kangaroo counselor, to start the events.

Finally Ruby blew her whistle and said, "We need two teams for tug-of-war. Let's split the group into halves so that each team has the same number of campers."

They counted off into two teams.

There were six campers on each team.

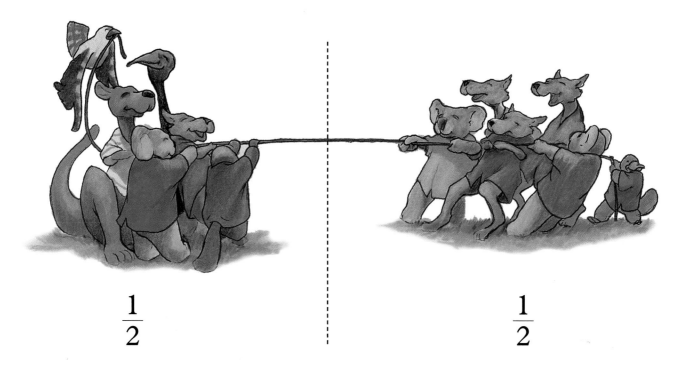

$\frac{1}{2}$ $\frac{1}{2}$

Each team had $\frac{1}{2}$ of the campers.

6 is $\frac{1}{2}$ of 12.

The campers tugged and grunted. And then they tugged some more. Both teams tugged very hard.

For a while, it looked like Kangaroo's team might win,
but with one giant tug the other team won instead.

Again, Ruby's whistle blew. "Now it's time for the swimming relay race. There are three lanes in the pool. Let's split into thirds so that we have an equal number of campers for each lane."

They counted off into three teams.

There were four campers on each team.

$\frac{1}{3}$ $\frac{1}{3}$ $\frac{1}{3}$

Each team had $\frac{1}{3}$ of the campers.

4 is $\frac{1}{3}$ of 12.

They kicked and splashed. And then they
kicked some more. All the campers swam
just as fast as they could.

This time, Kangaroo's team came in second.

Ruby blew her whistle once more. "Next we'll use our four canoes for a race on the lake. We need to split the group into fourths so that we have an equal number of campers in each canoe."

20

They counted off into four teams.

There were three campers on each team.

$\frac{1}{4}$ $\frac{1}{4}$ $\frac{1}{4}$ $\frac{1}{4}$

Each team had $\frac{1}{4}$ of the campers.

3 is $\frac{1}{4}$ of 12.

They paddled and pulled. And then they paddled some more. They paddled just as hard as they could.

They did their very best, but Kangaroo's team tied for last place. He wished that he could win just once.

"Cheer up, Kangaroo!" said Ruby. "You still have one more chance."

"Okay, everyone!" Ruby yelled, and then blew her whistle extra loud.

"Let's line up for the last event of the day—
the long jump. This time, you're on your own."

Each camper took a turn jumping. Finally, it was Kangaroo's turn.

Everyone yelled, "Jump, Kangaroo, jump!" He closed his eyes, sprang forward on his big hind legs, and jumped just as far as he could.

29

Kangaroo looked around in surprise as everyone cheered. Then, Ruby announced, "Kangaroo not only won the long jump, he set a new camp record."

"HOORAY FOR KANGAROO!"

FOR ADULTS AND KIDS

If you would like to have more fun with the math concepts presented in *Jump, Kangaroo, Jump!*, here are a few suggestions:

• Read the story with the child and talk about what is going on in each picture. Discuss why the group is splitting into a different number of teams for each competition.

• Ask questions throughout the story such as: "How many teams are needed?" "When the group is split so that each team has the same number of campers, how many are on each team?" "What fraction of the total number of campers is that?"

• Talk about fractions of groups. Give the child 12 bottle caps or buttons and pretend that each one is a camper. Encourage the child to divide the campers into 2 teams and tell what fraction of the group each team is. Try this with 3 and 4 teams.

• Talk about fractions of wholes. Each lane of the swimming pool is $\frac{1}{3}$ of the pool. Demonstrate other fractions of wholes by cutting a pizza or folding a piece of paper into halves, quarters, etc.

• Ask: "How many parts is the swimming pool divided into? What fraction do we call each part?" Then ask: "How many teams are the swimmers divided into? What fraction do we call each team? Why are the parts of the pool and the teams called by the same fraction?"

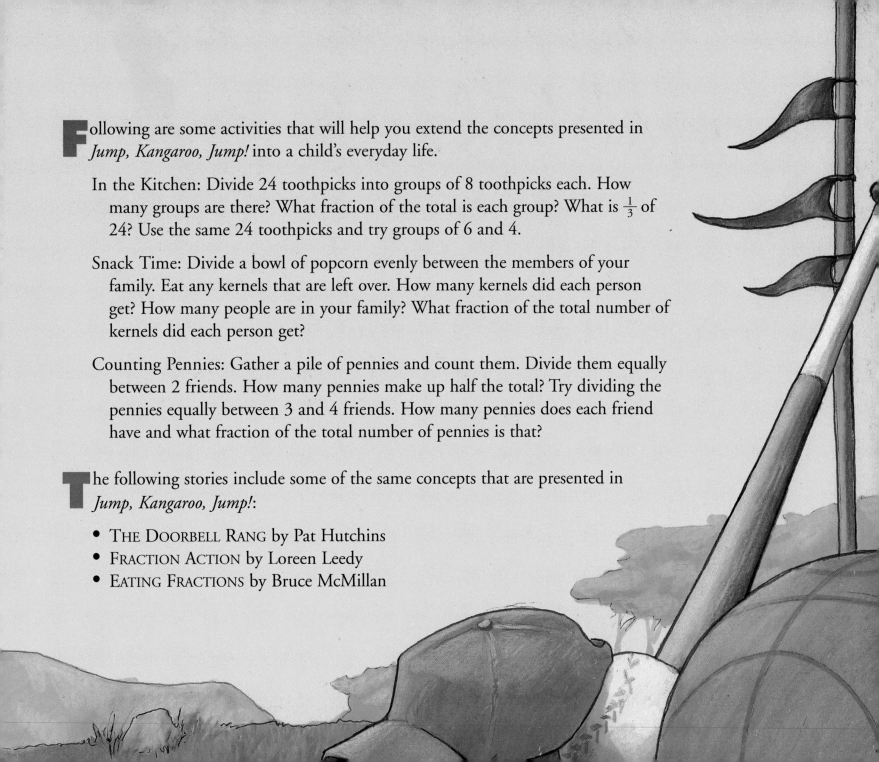

ollowing are some activities that will help you extend the concepts presented in *Jump, Kangaroo, Jump!* into a child's everyday life.

In the Kitchen: Divide 24 toothpicks into groups of 8 toothpicks each. How many groups are there? What fraction of the total is each group? What is $\frac{1}{3}$ of 24? Use the same 24 toothpicks and try groups of 6 and 4.

Snack Time: Divide a bowl of popcorn evenly between the members of your family. Eat any kernels that are left over. How many kernels did each person get? How many people are in your family? What fraction of the total number of kernels did each person get?

Counting Pennies: Gather a pile of pennies and count them. Divide them equally between 2 friends. How many pennies make up half the total? Try dividing the pennies equally between 3 and 4 friends. How many pennies does each friend have and what fraction of the total number of pennies is that?

he following stories include some of the same concepts that are presented in *Jump, Kangaroo, Jump!*:

- THE DOORBELL RANG by Pat Hutchins
- FRACTION ACTION by Loreen Leedy
- EATING FRACTIONS by Bruce McMillan